DANIIL KARABUT

The Secret Lives of Animals

"The greatness of a nation and its moral progress can be judged by the way its animals are treated."

Mahatma Gandhi

Contents

Foreword

Animals have always held a special place in the human imagination. From ancient cave paintings to modern-day wildlife documentaries, we have been captivated by the beauty, complexity, and diversity of the animal kingdom.

In "The Secret Lives of Animals," the author takes us on a journey of discovery, exploring the remarkable and fascinating world of animals, from the intelligence of elephants to the mysterious behavior of sharks and whales. Through detailed descriptions, personal anecdotes, and scientific insights, we come to appreciate the incredible diversity and complexity of the natural world, and the vital contributions that animals make to it.

But this book is not just a celebration of the animal kingdom. It is also a call to action. As we learn more about the secret lives of animals, we also come to understand the many threats that they face in their environments, from habitat loss to pollution to climate change. It is up to all of us to take action to protect and conserve animal populations, and to ensure a brighter future for the animal kingdom and for ourselves.

As you read this book, I invite you to immerse yourself in the secret lives of animals, to learn more about their behavior, communication, and ecological roles, and to be inspired to take action to protect and conserve the natural world. It is my hope that this book will not only entertain and educate, but also inspire readers to make a positive difference in the world around them.

Preface

As a lifelong animal lover, I have always been fascinated by the incredible diversity and complexity of the animal kingdom. From the smallest insects to the largest whales, animals have a remarkable ability to adapt to their environments, communicate with each other, and play critical roles in ecosystems around the world.

In "The Secret Lives of Animals," I invite readers to join me on a journey of discovery, as we explore the fascinating world of animal behavior, communication, and ecological roles. Through detailed descriptions and personal anecdotes, we come to appreciate the unique nature of each species, and the vital contributions they make to the natural world.

But this book is not just a celebration of the animal kingdom. It is also a call to action. As we learn more about the secret lives of animals, we also come to understand the many threats that they face in their environments, from habitat loss to pollution to climate change. It is up to all of us to take action to protect and conserve animal populations, and to ensure a brighter future for the animal kingdom and for ourselves.

I hope that this book will inspire readers to learn more about the animals that share our planet, and to take action to protect and conserve their populations. By working together, we can help to maintain the balance of ecosystems and preserve the remarkable diversity and beauty of the natural world for future generations.

Acknowledgement

I would like to express my heartfelt gratitude to all the people who have supported me throughout the process of writing this book.

First and foremost, I would like to thank my family and friends for their unwavering support and encouragement, and for always being there to lend a listening ear and a helping hand.

I am also deeply grateful to the researchers, conservationists, and wildlife experts who have generously shared their knowledge and expertise with me, and who have inspired me with their passion and dedication to protecting and conserving animal populations around the world.

Finally, I would like to express my gratitude to the readers of this book. It is my hope that this book will inspire you to learn more about the secret lives of animals and to take action to protect and conserve animal populations. Thank you for joining me on this journey of discovery, and for your commitment to preserving the natural world for future generations.

The Secret Social Lives of Elephants: How These Majestic Creatures Form Bonds and Communicate

Elephants are known for their intelligence, memory, and emotional complexity. But what many people may not realize is just how social and interconnected these animals truly are. In this chapter, we will explore the fascinating world of elephant socialization and communication.

Elephants live in family groups, led by a matriarch who has decades of experience and knowledge. These family units can be comprised of several generations of related females and their offspring, and are often composed of up to 20 individuals. Each family has its own unique set of behaviors, vocalizations, and social structures.

One of the most intriguing aspects of elephant socialization is their use of low-frequency vocalizations to communicate with one another over long distances. These rumbling sounds, which can be too low for human ears to detect, carry over several kilometers and can convey information about everything from food sources to danger warnings.

But elephants also communicate in more subtle ways. They use body language, such as ear flapping and trunk gestures, to express a wide range of emotions and intentions. They also engage in tactile communication, using their trunks to touch, greet, and comfort one another.

Beyond their own families, elephants also form social bonds with other elephants they encounter in their travels. These bonds can be based on shared experiences, such as migration or water scarcity, and can last for years or even decades.

In fact, elephants have been observed displaying empathy and compassion towards other elephants, even those outside of their family groups. In one well-known example, a group of elephants were seen mourning the death of a member of a different family, touching the body with their trunks and staying with it for several hours.

The social bonds between elephants serve many important functions. For example, they help elephants to protect one another from predators and find food and water sources. Older female elephants, in particular, are key to the survival of their families, as they possess valuable knowledge about migration routes and sources of food and water.

However, elephant socialization is not without its challenges. In areas where human activities have disrupted elephant habitats, conflicts can arise between elephants and humans. Elephants may raid crops, causing damage and economic losses for local farmers. In turn, humans may retaliate with violence, further exacerbating tensions and endangering both elephants and humans.

Despite these challenges, there are many efforts underway to protect and conserve elephant populations. These efforts range from anti-poaching campaigns to habitat restoration projects, and many involve working with local communities to find sustainable solutions that benefit both humans and elephants.

In this chapter, we will also explore some of the threats facing elephants today, from habitat loss to poaching for ivory. We will look at how conservationists and researchers are working to address these threats, and how the general public can get involved in elephant conservation efforts.

Unfortunately, the future of elephant populations is uncertain due to a number of threats. The loss of natural habitat due to human activities such as deforestation and agriculture has pushed elephants into smaller and smaller areas, increasing competition for resources and leading to conflicts with humans. Poaching for ivory is also a significant threat to elephant populations, with tens of thousands of elephants killed every year for their tusks.

In recent years, many conservationists and researchers have been working to address these threats and protect elephant populations. One approach has been to create protected areas and corridors that allow elephants to move freely between habitats. This can help to reduce human-elephant conflicts and provide safe havens for elephants to thrive.

Another approach is to engage with local communities to promote sustainable and wildlife-friendly practices. For example, programs have been implemented to provide alternative livelihoods for farmers living near elephant habitats, such as beekeeping or ecotourism, in order to reduce their reliance on crops that elephants may raid.

Conservationists and researchers are also working to combat poaching by increasing law enforcement efforts and reducing demand for ivory through public education campaigns. In addition, technologies such as GPS tracking and drones are being used to monitor elephant populations and track poaching activity.

The general public can also play a role in elephant conservation efforts. By supporting conservation organizations and ecotourism initiatives, individuals can help to fund research and conservation efforts on the ground. Choosing sustainable and wildlife-friendly products, such as palm oil-free products, can also help to reduce habitat loss and promote sustainable practices.

In conclusion, the social lives and communication strategies of elephants are fascinating and complex, but their survival is threatened by habitat loss and poaching. Conservation efforts are necessary to protect these majestic creatures and ensure their survival for future generations. By working together, we can help to preserve the secret lives of elephants and the many other incredible animals that share our planet.

The Secret Love Lives of Birds: The Intricate Courtship Rituals of Avian Species

Birds are a diverse and fascinating group of animals, known for their beautiful songs and stunning plumage. But birds are also famous for their elaborate courtship rituals, which can involve everything from colorful displays to complex vocalizations. In this chapter, we will explore the fascinating world of bird courtship and reproduction.

Birds use a variety of methods to attract and court mates. Some species rely on their colorful feathers, while others use complex vocalizations or impressive displays of athleticism. Male birds often go to great lengths to impress females, building intricate nests or performing intricate dances to win their affections.

One of the most well-known examples of bird courtship is the dance of the male bird of paradise. These birds have evolved a stunning array of colorful feathers, which they use to perform intricate dances and displays to attract females. These displays can be incredibly complex and unique, with each species having its own specific behaviors and movements.

But bird courtship is not only about impressing potential

mates, but also about choosing the right partner. Birds may engage in elaborate displays to assess the quality of potential mates, looking for signs of health, strength, and genetic fitness. In some cases, both males and females may have multiple partners, engaging in polygamous or polyandrous mating systems.

Once a pair of birds has bonded, they work together to build a nest and raise their offspring. Birds use a variety of techniques to incubate their eggs and care for their young, including sharing incubation duties, feeding and protecting their chicks, and teaching them to fly and hunt for food.

In some bird species, such as penguins and albatrosses, both parents play an active role in rearing their young, taking turns incubating eggs and feeding their chicks.

But bird courtship and reproduction are not without their challenges. Climate change, habitat loss, and pollution are all posing threats to bird populations around the world. Many bird species are facing declining populations and are at risk of extinction.

However, there are many efforts underway to protect and conserve bird populations. These efforts range from habitat restoration and anti-poaching campaigns to public education and outreach programs.

The general public can also play a role in bird conservation efforts. By supporting conservation organizations and initiatives, individuals can help to fund research and conservation efforts on the ground. They can also take steps to reduce their own impact on the environment, such as using fewer single-use plastics and supporting sustainable farming practices.

Overall, this chapter on the secret love lives of birds has revealed the fascinating and intricate world of bird courtship and reproduction. From the intricate displays and behaviors that birds use to attract mates to the complex relationships and social structures that exist between them, birds are truly remarkable creatures. However, their survival is threatened by human activities and climate change, making conservation efforts more important than ever.

The Secret Intelligence of Octopuses: How These Clever Cephalopods Navigate the Ocean and Solve Problems

Octopuses are among the most fascinating and intelligent creatures in the ocean. With their eight arms, remarkable camouflage abilities, and problem-solving skills, octopuses are truly masters of their environment. In this chapter, we will explore the amazing world of octopus intelligence and behavior.

Octopuses are known for their impressive problem-solving abilities. In laboratory experiments, octopuses have been shown to be able to navigate mazes, solve puzzles, and even use tools. They are also able to learn by observation, and can remember solutions to problems for weeks or even months.

One of the most remarkable aspects of octopus intelligence is their ability to camouflage themselves in their surroundings. Octopuses can change the color and texture of their skin to match their surroundings, making them almost invisible to predators. They can also use their camouflage abilities to sneak up on prey or to hide from danger.

In addition to their impressive problem-solving and camouflage skills, octopuses also have remarkable memories. They are able to recognize individuals and remember past experiences, which can help them to navigate their environment and avoid danger.

Octopuses are also known for their social behavior. While they are typically solitary creatures, they can communicate with one another through a range of visual and tactile signals. They may also engage in mating displays and other social behaviors.

However, octopuses face a number of threats in their environment. Overfishing, pollution, and climate change are all posing risks to octopus populations around the world. In addition, the intelligence and curiosity of octopuses has made them popular targets for scientific research and for the aquarium trade.

Efforts are underway to protect and conserve octopus populations. These efforts include habitat restoration and marine conservation initiatives, as well as efforts to reduce overfishing and pollution. In addition, researchers are working to better understand octopus behavior and intelligence, in order to better protect and conserve these remarkable creatures.

The general public can also play a role in octopus conservation efforts. By supporting marine conservation organizations and initiatives, individuals can help to fund research and conservation efforts on the ground. They can also take steps to reduce their own impact on the environment, such as reducing single-use plastics and supporting sustainable fishing practices.

Overall, this chapter on the secret intelligence of octopuses has revealed the fascinating and complex world of these remarkable creatures. With their impressive problem-solving skills, remarkable camouflage abilities, and social behavior, octopuses are truly amazing creatures. However, they face many threats in their environment, making conservation efforts more important than ever.

The Secret Life of Bees: The Fascinating World of These Important Pollinators

Bees are one of the most important pollinators in the world, playing a critical role in the pollination of many of our food crops and wild plants. But bees are also fascinating creatures in their own right, with complex social structures, unique communication methods, and impressive abilities. In this chapter, we will explore the secret life of bees and their importance to our world.

Bees live in complex social structures, with each colony consisting of a queen bee, male drones, and female worker bees. The queen bee is responsible for laying eggs, while the worker bees are responsible for tasks such as collecting nectar and pollen, building the hive, and caring for the young. The drones play a role in reproduction, mating with the queen bee to produce new offspring.

One of the most impressive aspects of bee behavior is their communication methods. Bees use a unique dance language, known as the waggle dance, to communicate with one another about the location of food sources. This dance involves a series of movements that convey information about the direction and distance of the food source, as well as its quality.

In addition to their communication abilities, bees also have impressive navigational skills. They are able to navigate using the sun and the Earth's magnetic field, allowing them to travel long distances in search of food sources.

But bees are facing many threats in their environment. Pesticides, habitat loss, and climate change are all posing risks to bee populations around the world. This is concerning, as bees play a critical role in pollinating many of our food crops and wild plants. Without bees, our world would be a much less productive and diverse place.

Efforts are underway to protect and conserve bee populations. These efforts include habitat restoration and conservation initiatives, as well as campaigns to reduce the use of pesticides and promote sustainable farming practices. In addition, researchers are working to better understand bee behavior and the threats facing bee populations, in order to develop effective conservation strategies.

The general public can also play a role in bee conservation efforts. By supporting conservation organizations and initiatives, individuals can help to fund research and conservation efforts on the ground. They can also take steps to reduce their own impact on the environment, such as using fewer pesticides and supporting sustainable farming practices.

Overall, this chapter on the secret life of bees has revealed the fascinating world of these important pollinators. With their complex social structures, unique communication methods, and impressive navigational skills, bees are truly remarkable creatures. However, they face many threats in their environment, making conservation efforts more important than ever.

The Secret World of Ants: How These Tiny Insects Build Complex Societies and Work Together to Survive

Ants are among the most numerous and diverse animals on the planet, with an estimated 10 quadrillion ants living on Earth. But ants are more than just a common sight in our backyards and gardens – they are also incredibly fascinating creatures, known for their complex social structures and impressive teamwork. In this chapter, we will explore the secret world of ants and how they work together to survive.

Ants live in highly organized societies, with each individual ant having a specific role to play in the colony. Queen ants are responsible for laying eggs, while worker ants are responsible for tasks such as gathering food, caring for the young, and defending the colony. Some ant species also have soldier ants, which are responsible for defending the colony against predators.

One of the most remarkable aspects of ant behavior is their ability to work together to solve complex problems. Ants use pheromones to communicate with one another, laying down scent trails to mark the location of food sources or potential threats. They also work together to build elaborate nests, using materials such as dirt, leaves, and twigs to create intricate structures.

Ants are also known for their impressive strength and endurance. Some ant species can carry objects many times their own weight, while others are able to travel long distances in search of food sources.

But ants are facing many threats in their environment. Habitat loss, climate change, and pesticide use are all posing risks to ant populations around the world. This is concerning, as ants play a critical role in maintaining ecosystems and providing important ecological services such as soil aeration and pest control.

Efforts are underway to protect and conserve ant populations. These efforts include habitat restoration and conservation initiatives, as well as campaigns to reduce the use of pesticides and promote sustainable farming practices. In addition, researchers are working to better understand ant behavior and the threats facing ant populations, in order to develop effective conservation strategies.

The general public can also play a role in ant conservation efforts. By supporting conservation organizations and initiatives, individuals can help to fund research and conservation efforts on the ground. They can also take steps to reduce their own impact on the environment, such as using fewer pesticides and supporting sustainable farming practices.

Overall, this chapter on the secret world of ants has revealed the fascinating and complex nature of these tiny insects. With their highly organized societies, impressive teamwork, and strength, ants are truly remarkable creatures. However, they face many threats in their environment, making conservation efforts more important than ever.

The Secret Lives of Dolphins: How These Intelligent Marine Mammals Navigate the Ocean and Communicate with One Another

Dolphins are some of the most intelligent and charismatic creatures in the ocean, known for their playful behavior, impressive acrobatics, and remarkable intelligence. In this chapter, we will explore the fascinating world of dolphin behavior and how these intelligent marine mammals navigate the ocean and communicate with one another.

Dolphins are highly social creatures, living in groups called pods. These pods can range in size from just a few individuals to hundreds of dolphins. Within each pod, dolphins form complex social bonds, engaging in behaviors such as grooming and playing with one another.

One of the most remarkable aspects of dolphin behavior is their communication abilities. Dolphins use a range of vocalizations, including clicks, whistles, and squeaks, to communicate with one another. They are also able to use echolocation, bouncing sound waves off objects in their environment to navigate and find prey.

Dolphins are also known for their intelligence and problem-solving abilities. In laboratory experiments, dolphins have been shown to be able to solve complex problems, such as opening latches and using tools. They are also able to learn by observation and can remember solutions to problems for long periods of time.

But dolphins are facing many threats in their environment. Overfishing, pollution, and climate change are all posing risks to dolphin populations around the world. This is concerning, as dolphins play a critical role in maintaining the balance of ocean ecosystems and are an important part of many coastal communities.

Efforts are underway to protect and conserve dolphin populations. These efforts include habitat restoration and marine conservation initiatives, as well as campaigns to reduce overfishing and pollution. In addition, researchers are working to better understand dolphin behavior and the threats facing dolphin populations, in order to develop effective conservation strategies.

The general public can also play a role in dolphin conservation efforts. By supporting conservation organizations and initiatives, individuals can help to fund research and conservation efforts on the ground. They can also take steps to reduce their own impact on the environment, such as reducing single-use plastics and supporting sustainable fishing practices.

Overall, this chapter on the secret lives of dolphins has revealed the fascinating and complex world of these intelligent marine mammals. With their complex social behavior, remarkable communication abilities, and problem-solving skills, dolphins are truly remarkable creatures. However, they face many threats in their environment, making conservation efforts more important than ever.

The Secret Lives of Bats: How These Misunderstood Creatures Navigate the Night and Contribute to Ecosystems

Bats are often misunderstood creatures, often portrayed as spooky or even dangerous. But in reality, bats are fascinating animals with remarkable abilities and important roles in ecosystems around the world. In this chapter, we will explore the secret lives of bats and their contributions to the natural world.

Bats are the only mammals capable of sustained flight, thanks to their wings, which are made up of thin layers of skin stretched over elongated finger bones. With their ability to fly and echolocation, bats are uniquely equipped to navigate the night sky and hunt for insects and other prey.

In addition to their impressive flight abilities, bats are also important pollinators and seed dispersers. Some bat species are responsible for pollinating many of our food crops, while others play a critical role in spreading seeds throughout forests and other ecosystems.

Bats also play an important role in controlling insect populations. Some bat species can consume thousands of insects in a single night, making them an important natural form of pest control.

But bats are facing many threats in their environment. Habitat loss, climate change, and disease are all posing risks to bat populations around the world. This is concerning, as bats play a critical role in maintaining the balance of ecosystems and are an important part of many cultural traditions.

Efforts are underway to protect and conserve bat populations. These efforts include habitat restoration and conservation initiatives, as well as campaigns to reduce the use of pesticides and promote sustainable farming practices. In addition, researchers are working to better understand bat behavior and the threats facing bat populations, in order to develop effective conservation strategies.

The general public can also play a role in bat conservation efforts. By supporting conservation organizations and initiatives, individuals can help to fund research and conservation efforts on the ground. They can also take steps to reduce their own impact on the environment, such as reducing single-use plastics and supporting sustainable farming practices.

Overall, this chapter on the secret lives of bats has revealed the fascinating and complex nature of these often-misunderstood creatures. With their impressive flight abilities, important ecological roles, and critical contributions to many cultural traditions, bats are truly remarkable animals. However, they face many threats in their environment, making conservation efforts more important than ever.

The Secret Lives of Wolves: How These Iconic Predators Form Packs and Navigate Their Environments

Wolves are some of the most iconic and fascinating predators on the planet, known for their intelligence, social behavior, and impressive hunting abilities. In this chapter, we will explore the secret lives of wolves and how they form packs and navigate their environments.

Wolves are highly social creatures, living in packs that can range in size from just a few individuals to over a dozen. Within each pack, wolves form complex social hierarchies, with dominant individuals taking on leadership roles and others following their lead.

One of the most remarkable aspects of wolf behavior is their ability to work together to take down large prey. Wolves use coordinated hunting strategies, such as surrounding their prey and taking turns attacking, to bring down animals such as elk, moose, and bison.

Wolves are also known for their impressive sense of smell, which they use to navigate their environments and locate prey. Their sense of smell is so acute that they are able to detect scents from up to a mile away.

But wolves are facing many threats in their environment. Habitat loss, hunting, and persecution are all posing risks to wolf populations around the world. This is concerning, as wolves play a critical role in maintaining the balance of ecosystems and are an important part of many cultural traditions.

Efforts are underway to protect and conserve wolf populations. These efforts include habitat restoration and conservation initiatives, as well as campaigns to reduce hunting and persecution. In addition, researchers are working to better understand wolf behavior and the threats facing wolf populations, in order to develop effective conservation strategies.

The general public can also play a role in wolf conservation efforts. By supporting conservation organizations and initiatives, individuals can help to fund research and conservation efforts on the ground. They can also take steps to reduce their own impact on the environment, such as supporting sustainable land use practices.

Overall, this chapter on the secret lives of wolves has revealed the fascinating and complex nature of these iconic predators. With their highly organized social behavior, impressive hunting abilities, and acute sense of smell, wolves are truly remarkable creatures. However, they face many threats in their environment, making conservation efforts more important than ever.

The Secret Lives of Penguins: How These Flightless Birds Survive in Some of the Harshest Environments on Earth

Penguins are some of the most unique and beloved birds on the planet, known for their waddling gait, striking black and white feathers, and their ability to survive in some of the most extreme environments on earth. In this chapter, we will explore the secret lives of penguins and how they navigate their harsh environments.

Penguins are flightless birds that are found primarily in the Southern Hemisphere. There are 18 species of penguins, ranging in size from the tiny fairy penguin to the emperor penguin, which can grow up to four feet tall.

One of the most remarkable aspects of penguin behavior is their ability to survive in some of the harshest environments on earth. Penguins can survive in temperatures as low as -40 degrees Fahrenheit and can withstand winds of up to 100 miles per hour. They do this by huddling together for warmth, with individuals rotating positions to ensure that everyone stays warm.

Penguins are also excellent swimmers, with some species able to dive to depths of over 500 feet. They use their wings as flippers to propel themselves through the water and catch fish and other prey.

But penguins are facing many threats in their environment. Climate change, pollution, and overfishing are all posing risks to penguin populations around the world. This is concerning, as penguins play a critical role in maintaining the balance of ocean ecosystems and are an important part of many cultural traditions.

Efforts are underway to protect and conserve penguin populations. These efforts include marine conservation initiatives, campaigns to reduce overfishing and pollution, and efforts to reduce carbon emissions and slow the effects of climate change. In addition, researchers are working to better understand penguin behavior and the threats facing penguin populations, in order to develop effective conservation strategies.

The general public can also play a role in penguin conservation efforts. By supporting conservation organizations and initiatives, individuals can help to fund research and conservation efforts on the ground. They can also take steps to reduce their own impact on the environment, such as reducing single-use plastics and supporting sustainable fishing practices.

Overall, this chapter on the secret lives of penguins has revealed the remarkable and unique nature of these flightless birds. With their ability to survive in extreme environments, their impressive swimming abilities, and their critical ecological role, penguins are truly remarkable creatures. However, they face many threats in their environment, making conservation efforts more important than ever.

The Secret Lives of Chimpanzees: How These Close Relatives of Humans Use Tools, Communicate, and Form Complex Societies

Chimpanzees are one of the closest living relatives of humans, sharing more than 98% of our DNA. In this chapter, we will explore the fascinating world of chimpanzees and their secret lives, including their use of tools, complex social behaviors, and communication methods.

One of the most remarkable aspects of chimpanzee behavior is their use of tools. Chimpanzees use sticks, rocks, and other objects to access food, create shelter, and defend themselves against predators. They have also been observed using tools to fish for termites and ants, and to crack open nuts.

Chimpanzees also exhibit complex social behaviors, living in large groups or communities that can number up to 150 individuals. Within these communities, chimpanzees form complex social hierarchies, with dominant individuals taking on leadership roles and others following their lead.

Chimpanzees also communicate with each other in a variety of ways, including through vocalizations, body language, and facial expressions. They have been observed using gestures to communicate and have even been shown to understand human sign language.

But chimpanzees are facing many threats in their environment, including habitat loss, hunting, and the illegal pet trade. This is concerning, as chimpanzees play a critical role in maintaining the balance of ecosystems and are an important part of many cultural traditions.

Efforts are underway to protect and conserve chimpanzee populations. These efforts include habitat restoration and conservation initiatives, as well as campaigns to reduce hunting and the illegal pet trade. In addition, researchers are working to better understand chimpanzee behavior and the threats facing chimpanzee populations, in order to develop effective conservation strategies.

The general public can also play a role in chimpanzee conservation efforts. By supporting conservation organizations and initiatives, individuals can help to fund research and conservation efforts on the ground. They can also take steps to reduce their own impact on the environment, such as supporting sustainable land use practices.

Overall, this chapter on the secret lives of chimpanzees has revealed the remarkable and complex nature of these close relatives of humans. With their use of tools, complex social behaviors, and unique communication methods, chimpanzees are truly fascinating creatures. However, they face many threats in their environment, making conservation efforts more important than ever.

The Secret Lives of Sharks: How These Powerful Predators Navigate the Ocean and Play Critical Roles in Ecosystems

Sharks are some of the most fascinating and misunderstood creatures in the ocean, known for their powerful hunting abilities and unique adaptations. In this chapter, we will explore the secret lives of sharks and their contributions to the natural world.

Sharks are a diverse group of fish that are found in oceans around the world. There are over 500 known species of sharks, ranging in size from the tiny pygmy shark to the massive whale shark, which can grow up to 40 feet long.

One of the most remarkable aspects of shark behavior is their ability to navigate the ocean with incredible accuracy. Sharks use their sense of smell and the Earth's magnetic field to navigate long distances and find their prey. They can also detect the electric fields produced by other animals in the water, allowing them to locate hidden prey.

Sharks are also powerful hunters, with some species able to swim at speeds of up to 60 miles per hour. They use their sharp teeth and powerful jaws to catch and consume a wide range of prey, including fish, squid, and marine mammals.

But sharks are facing many threats in their environment, including overfishing, habitat loss, and pollution. This is concerning, as sharks play a critical role in maintaining the balance of ocean ecosystems and are an important part of many cultural traditions.

Efforts are underway to protect and conserve shark populations. These efforts include marine conservation initiatives, campaigns to reduce overfishing and pollution, and efforts to reduce carbon emissions and slow the effects of climate change. In addition, researchers are working to better understand shark behavior and the threats facing shark populations, in order to develop effective conservation strategies.

The general public can also play a role in shark conservation efforts. By supporting conservation organizations and initiatives, individuals can help to fund research and conservation efforts on the ground. They can also take steps to reduce their own impact on the environment, such as reducing single-use plastics and supporting sustainable fishing practices.

Overall, the secret lives of sharks reveal the remarkable and unique nature of these powerful predators. With their ability to navigate the ocean with incredible accuracy and their critical ecological role, sharks are truly remarkable creatures. However, they face many threats in their environment, making conservation efforts more important than ever. By protecting and conserving shark populations, we can help to maintain the balance of ocean ecosystems and preserve these fascinating creatures for future generations.

The Secret Lives of Sea Turtles: How These Ancient Creatures Navigate the Ocean and Contribute to Ecosystems around the World

Sea turtles are some of the most ancient and majestic creatures on the planet, having been swimming the world's oceans for millions of years. In this chapter, we will explore the secret lives of sea turtles and their contributions to the natural world.

There are seven known species of sea turtles, each with their own unique characteristics and behaviors. Sea turtles are found in all of the world's oceans, from the shallow waters of the tropics to the deep sea.

One of the most remarkable aspects of sea turtle behavior is their ability to navigate the ocean with incredible accuracy. Sea turtles use the Earth's magnetic field and the position of the sun and stars to navigate long distances and return to their natal beaches to lay their eggs. They can also hold their breath for up to five hours, allowing them to remain underwater for extended periods of time.

Sea turtles play a critical role in maintaining the balance of ocean ecosystems. They are herbivores, feeding on seagrass and other marine vegetation, and serve as an important food source for predators such as sharks and crocodiles. They also help to maintain healthy coral reefs by grazing on algae that can smother corals.

But sea turtles are facing many threats in their environment, including habitat loss, pollution, and overfishing. This is concerning, as sea turtles play a critical role in maintaining the balance of ocean ecosystems and are an important part of many cultural traditions.

Efforts are underway to protect and conserve sea turtle populations. These efforts include marine conservation initiatives, campaigns to reduce plastic pollution and other forms of marine debris, and efforts to reduce carbon emissions and slow the effects of climate change. In addition, researchers are working to better understand sea turtle behavior and the threats facing sea turtle populations, in order to develop effective conservation strategies.

The general public can also play a role in sea turtle conservation efforts. By supporting conservation organizations and initiatives, individuals can help to fund research and conservation efforts on the ground. They can also take steps to reduce their own impact on the environment, such as reducing single-use plastics and supporting sustainable fishing practices.

Overall, the secret lives of sea turtles reveal the remarkable and ancient nature of these majestic creatures. With their ability to navigate the ocean with incredible accuracy and their critical ecological role, sea turtles are truly remarkable creatures. However, they face many threats in their environment, making conservation efforts more important than ever. By protecting and conserving sea turtle populations, we can help to maintain the balance of ocean ecosystems and preserve these fascinating creatures for future generations.

The Secret Lives of Whales: How These Majestic Marine Mammals Communicate, Navigate the Ocean, and Play Critical Roles in Ecosystems

Whales are some of the largest and most magnificent creatures in the ocean, known for their impressive size, songs, and unique adaptations. In this chapter, we will explore the secret lives of whales and their contributions to the natural world.

There are over 80 known species of whales, ranging in size from the small dwarf sperm whale to the massive blue whale, which can grow up to 100 feet long. Whales are found in all of the world's oceans and play a critical role in maintaining the balance of marine ecosystems.

One of the most remarkable aspects of whale behavior is their ability to communicate with each other over long distances. Whales use a variety of vocalizations, including songs, clicks, and whistles, to communicate with each other and navigate the ocean. Some species of whales, such as humpbacks, are known for their complex songs, which can last for up to 20 minutes and are thought to play a role in mating and social behavior.

Whales are also powerful swimmers, able to dive to depths of up to 2,000 feet and stay underwater for up to two hours. They use their flippers and tails to propel themselves through the water and navigate the ocean.

But whales are facing many threats in their environment, including overfishing, habitat loss, and pollution. This is concerning, as whales play a critical role in maintaining the balance of ocean ecosystems and are an important part of many cultural traditions.

Efforts are underway to protect and conserve whale populations. These efforts include marine conservation initiatives, campaigns to reduce overfishing and pollution, and efforts to reduce carbon emissions and slow the effects of climate change. In addition, researchers are working to better understand whale behavior and the threats facing whale populations, in order to develop effective conservation strategies.

The general public can also play a role in whale conservation efforts. By supporting conservation organizations and initiatives, individuals can help to fund research and conservation efforts on the ground. They can also take steps to reduce their own impact on the environment, such as reducing single-use plastics and supporting sustainable fishing practices.

Overall, the secret lives of whales reveal the remarkable and unique nature of these majestic marine mammals. With their ability to communicate over long distances, navigate the ocean with incredible accuracy, and play a critical ecological role, whales are truly remarkable creatures. However, they face many threats in their environment, making conservation efforts more important than ever. By protecting and conserving whale populations, we can help to maintain the balance of ocean ecosystems and preserve these fascinating creatures for future generations.

The Secret Lives of Birds of Prey: How These Magnificent Raptors Hunt, Communicate, and Play Critical Roles in Ecosystems

Birds of prey, also known as raptors, are a group of birds that includes hawks, eagles, owls, falcons, and vultures. These birds are known for their sharp talons, powerful beaks, and impressive hunting abilities. In this chapter, we will explore the secret lives of birds of prey and their contributions to the natural world.

One of the most remarkable aspects of bird of prey behavior is their hunting abilities. These birds have sharp talons and beaks, which they use to catch and kill their prey. They have excellent eyesight and can see small prey from great distances, allowing them to hunt efficiently.

Birds of prey also communicate with each other in a variety of ways, including through vocalizations and body language. They use calls to communicate with their mates, establish territories, and warn other birds of danger.

Birds of prey play a critical role in maintaining the balance of ecosystems. They are top predators, hunting and consuming other animals, which helps to keep populations in check. They also provide important ecological services, such as scavenging and cleaning up dead animals, which helps to prevent the spread of disease.

But birds of prey are facing many threats in their environment, including habitat loss, hunting, and pollution. This is concerning, as birds of prey play a critical role in maintaining the balance of ecosystems and are an important part of many cultural traditions.

Efforts are underway to protect and conserve bird of prey populations. These efforts include habitat restoration and conservation initiatives, campaigns to reduce hunting and the illegal trade of birds of prey, and efforts to reduce pollution and slow the effects of climate change. In addition, researchers are working to better understand bird of prey behavior and the threats facing bird of prey populations, in order to develop effective conservation strategies.

The general public can also play a role in bird of prey conservation efforts. By supporting conservation organizations and initiatives, individuals can help to fund research and conservation efforts on the ground. They can also take steps to reduce their own impact on the environment, such as reducing single-use plastics and supporting sustainable land use practices.

Overall, the secret lives of birds of prey reveal the remarkable and unique nature of these magnificent raptors. With their impressive hunting abilities, communication methods, and critical ecological role, birds of prey are truly fascinating creatures. However, they face many threats in their environment, making conservation efforts more important than ever. By protecting and conserving bird of prey populations, we can help to maintain the balance of ecosystems and preserve these magnificent creatures for future generations.

The Secret Lives of Insects: How These Tiny Creatures Shape Ecosystems and Impact Human Society

Insects are some of the most diverse and numerous creatures on the planet, with over a million known species. These tiny creatures play a critical role in maintaining the balance of ecosystems and impact human society in many ways. In this chapter, we will explore the secret lives of insects and their contributions to the natural world.

Insects play a critical role in pollination, with many species serving as important pollinators for crops and wild plants. They also serve as food for a wide range of other animals, from birds to bats to larger mammals.

But insects are also known for their unique behaviors, such as the complex social behavior of ants and the incredible metamorphosis of butterflies. They communicate with each other through chemical signals and use a wide range of adaptations, from camouflage to mimicry, to survive in their environments.

However, insects also impact human society in many ways. Some insects, such as mosquitoes, can spread disease and cause significant health problems. Other insects, such as termites and cockroaches, can cause damage to buildings and homes.

Still, other insects, such as honeybees, play a critical role in agriculture, pollinating crops and providing honey and beeswax.

Efforts are underway to protect and conserve insect populations, including campaigns to reduce pesticide use and protect habitat for pollinators. In addition, researchers are working to better understand insect behavior and the impact of human activities on insect populations.

The general public can also play a role in insect conservation efforts. By supporting conservation organizations and initiatives, individuals can help to fund research and conservation efforts on the ground. They can also take steps to reduce their own impact on the environment, such as reducing pesticide use and planting native plants to provide habitat for pollinators.

Overall, the secret lives of insects reveal the remarkable and unique nature of these tiny creatures. With their critical ecological roles, fascinating behaviors, and impact on human society, insects are truly remarkable creatures. However, they face many threats in their environment, making conservation efforts more important than ever. By protecting and conserving insect populations, we can help to maintain the balance of ecosystems and preserve these fascinating creatures for future generations.

Conclusion

In this book, we have explored the secret lives of animals, from the incredible intelligence of elephants to the mysterious behavior of sharks, sea turtles, and whales. We have seen how animals communicate, navigate, and play critical roles in ecosystems around the world.

Through our exploration of the animal kingdom, we have come to appreciate the remarkable and unique nature of each species, and the vital contributions they make to the natural world. However, we have also seen how animals are facing many threats in their environment, from habitat loss to pollution to climate change.

Despite the challenges facing the animal kingdom, there is hope. Conservation efforts are underway to protect and conserve animal populations, and researchers are working to better understand animal behavior and the threats they face. The general public can also play a role in animal conservation efforts, through supporting conservation organizations and taking steps to reduce their own impact on the environment.

Ultimately, the secret lives of animals reveal the incredible diversity and complexity of the natural world, and remind us of the importance of protecting and preserving these remarkable creatures for future generations. By taking action to conserve and protect animal populations, we can help to maintain the balance of ecosystems and ensure a brighter future for the animal kingdom and for ourselves.

Afterword

Writing this book has been an incredible journey of discovery, and I hope that readers have found it as fascinating and enlightening as I have. The animal kingdom is truly remarkable, and the more we learn about the secret lives of animals, the more we come to appreciate the complexity and beauty of the natural world.

As we have seen throughout this book, animals are facing many threats in their environment, from habitat loss to climate change to overfishing and poaching. However, there is hope. Conservation efforts are underway around the world to protect and conserve animal populations, and researchers are working tirelessly to better understand animal behavior and the threats they face.

It is also important to recognize the role that individuals can play in animal conservation efforts. By making small changes in our own lives, such as reducing our carbon footprint and supporting sustainable practices, we can help to reduce our impact on the environment and protect the animals that call it home. Supporting conservation organizations and initiatives is also a powerful way to make a difference.

As we move forward, it is my hope that this book will inspire readers to learn more about the animals that share our planet, and to take action to protect and conserve their populations. By working together, we can help to ensure a brighter future for the animal kingdom and for ourselves.